What's this, tucked way up under a roof? It's brown and furry. It opens its mouth—full of sharp teeth. It's a bat! Will it bite you and suck your blood? Will it chase you and get tangled in your hair?

Bats! At night they come out and fly around in the dark. They are strange and scary, with their fluttery wings and big ears.

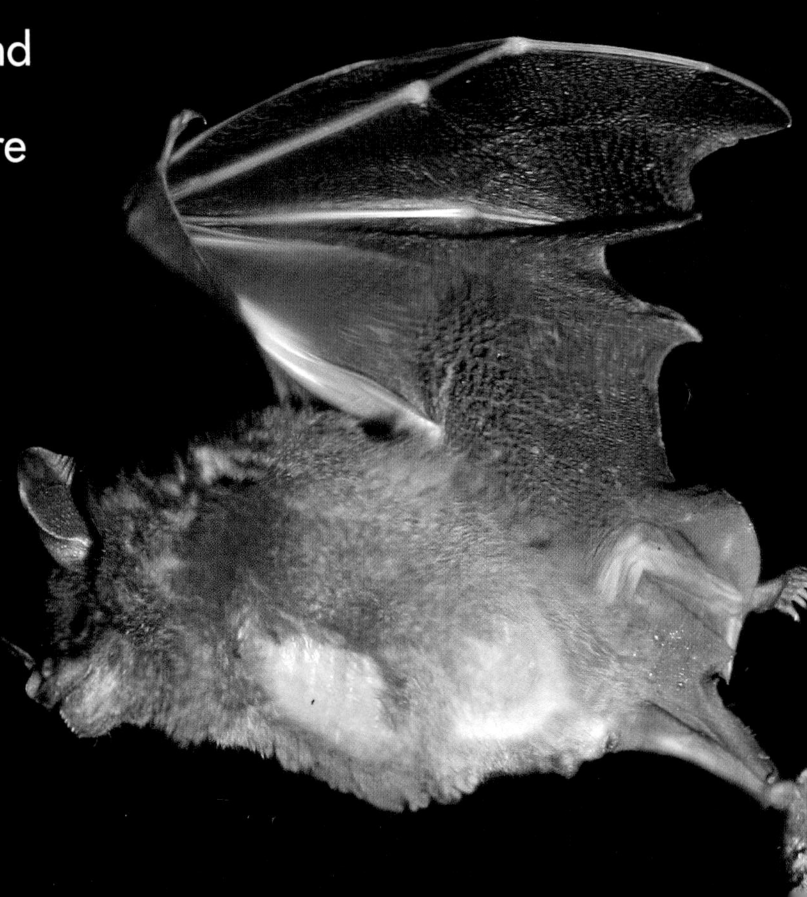

They hang upside down and hide during the day. Do bats make you shiver with fright? If you knew what bats are really like, you wouldn't be afraid.

I USED to be AFRAID of BATS,
but now I know . . .

Bats are amazing fliers. They dive, dart, and turn quickly. A bat's wings are not like the wings of birds. They do not have feathers.

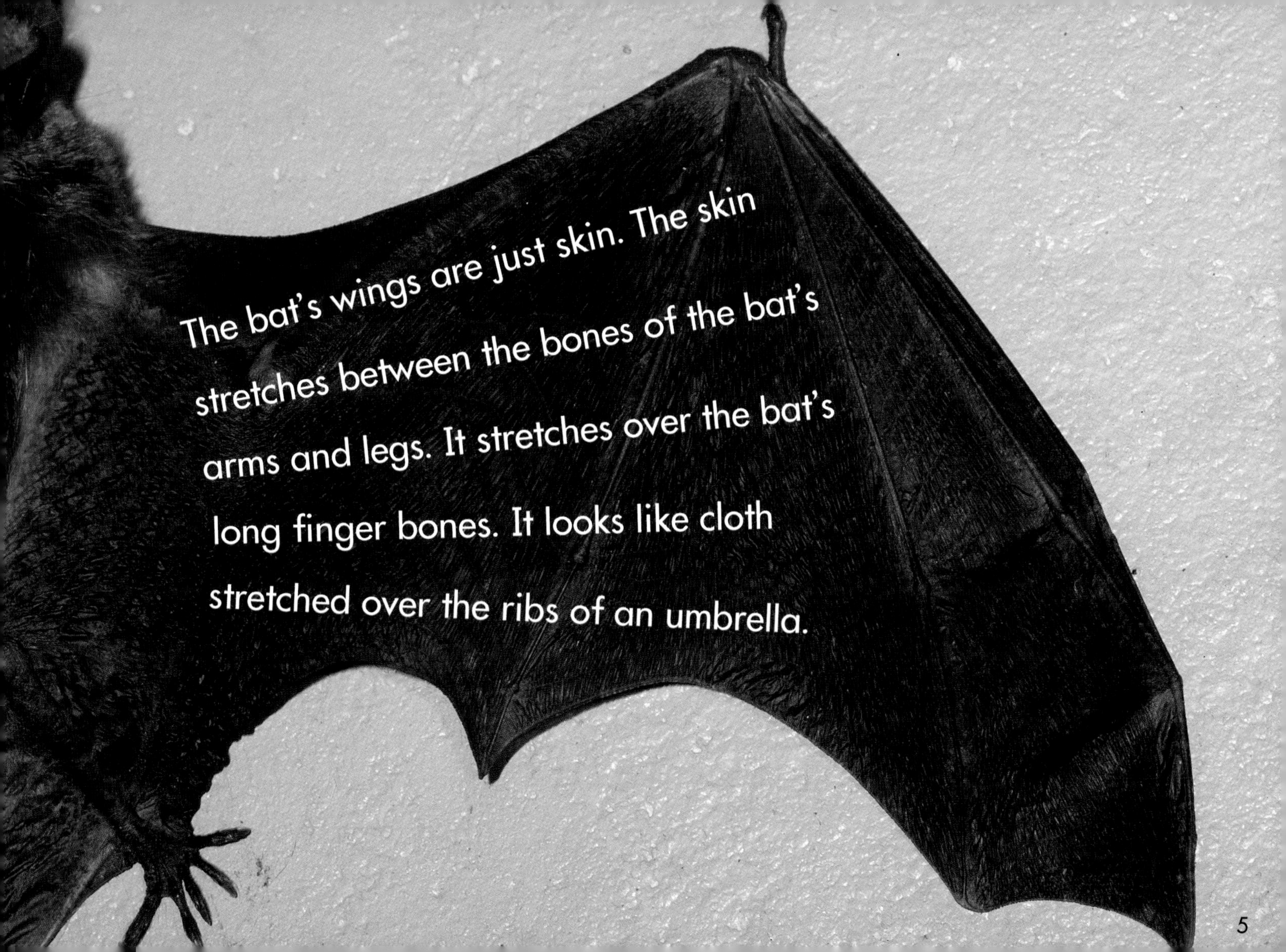

The bat's wings are just skin. The skin stretches between the bones of the bat's arms and legs. It stretches over the bat's long finger bones. It looks like cloth stretched over the ribs of an umbrella.

I USED to be AFRAID of BATS,
but now I know . . .

Like most bats, this big brown bat eats flying insects. It has to be a good flier. The bat zooms around, snatching its dinner right out of the air.

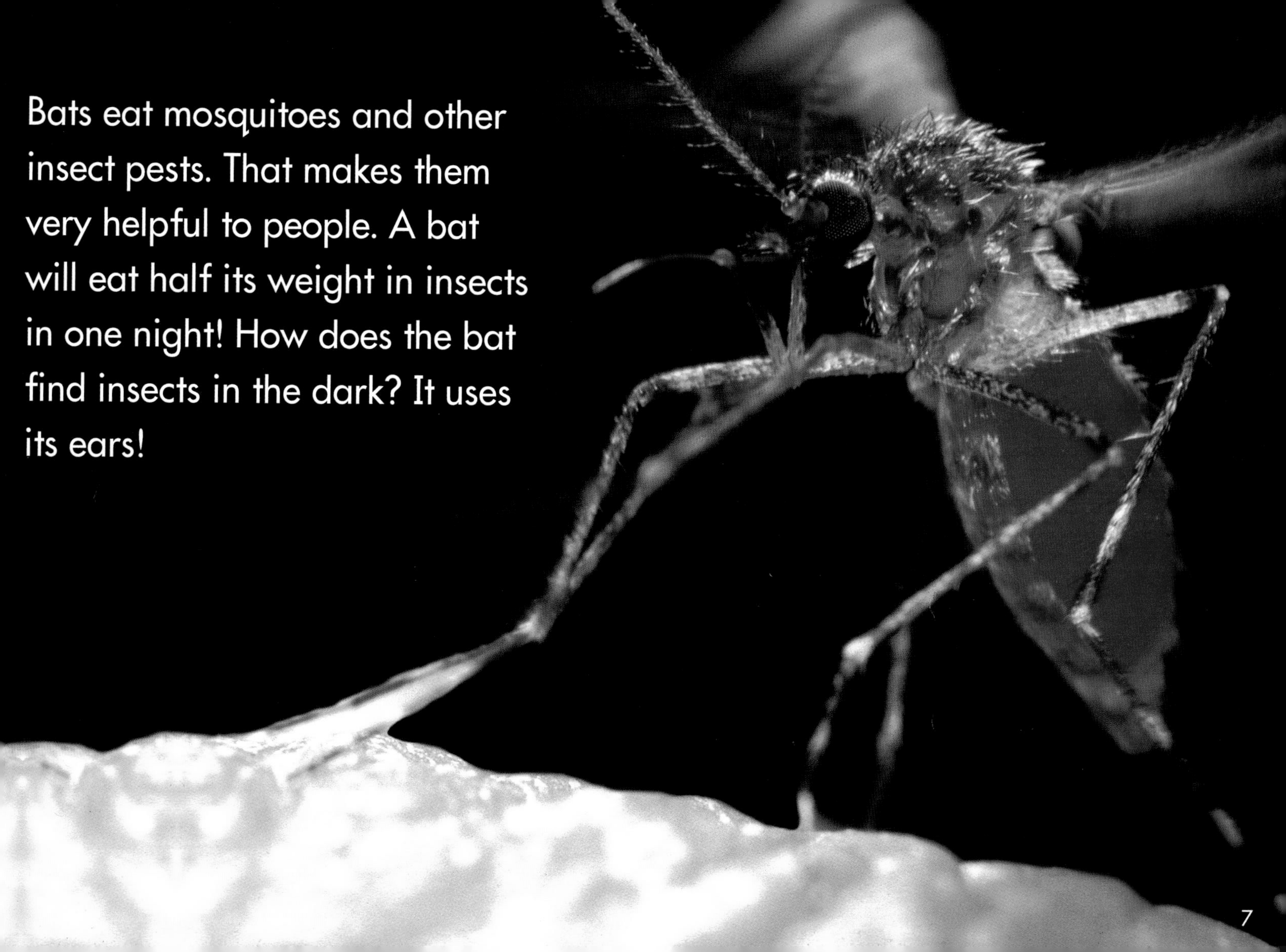

Bats eat mosquitoes and other insect pests. That makes them very helpful to people. A bat will eat half its weight in insects in one night! How does the bat find insects in the dark? It uses its ears!

I USED to be AFRAID of BATS,
but now I know . . .

The bat squeaks as it flies. You can't hear the squeaks, but bats can. The sounds travel through the air. They bounce off objects, making echoes.

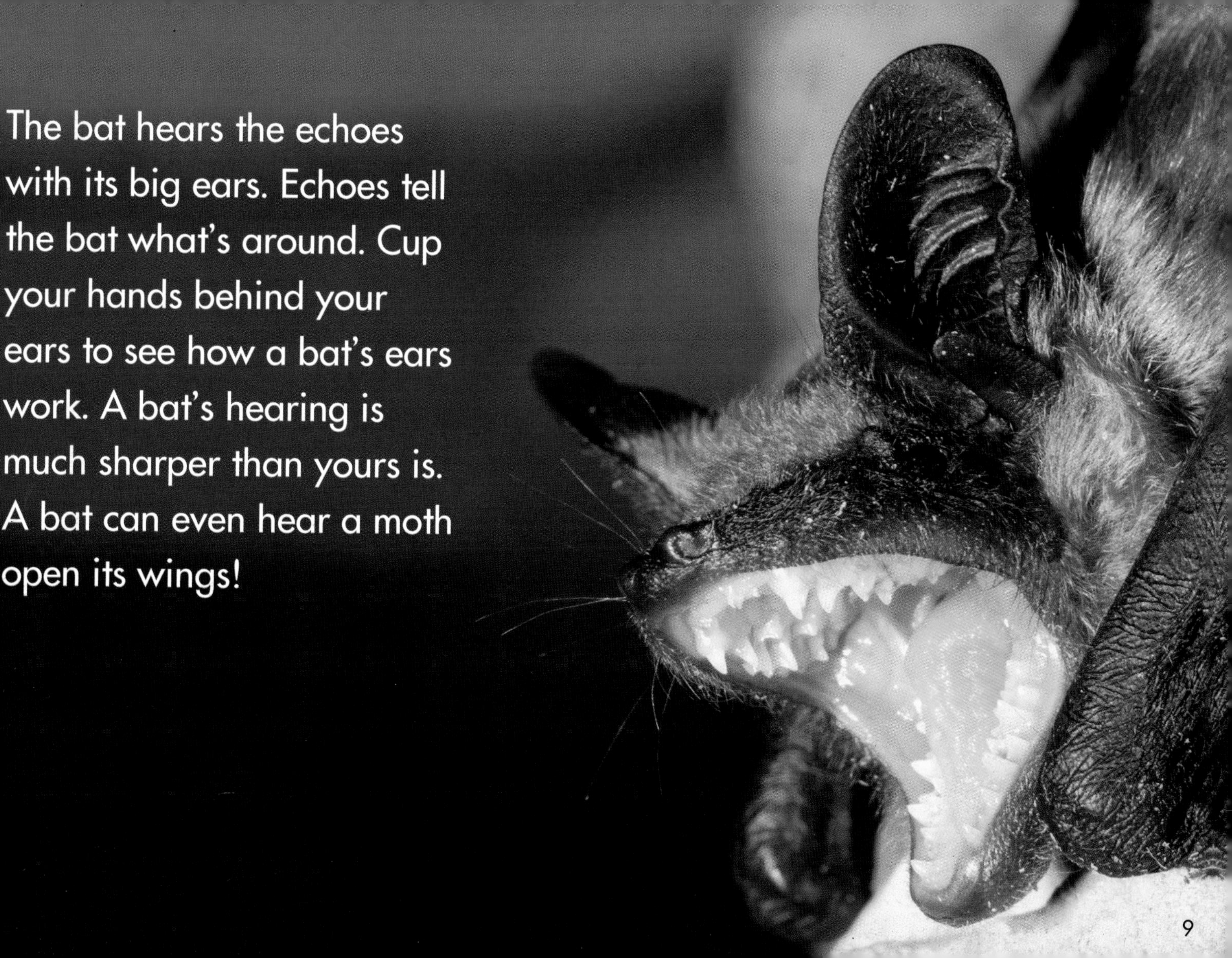

The bat hears the echoes with its big ears. Echoes tell the bat what's around. Cup your hands behind your ears to see how a bat's ears work. A bat's hearing is much sharper than yours is. A bat can even hear a moth open its wings!

Not all bats eat insects. These little bats like bananas. Bats that eat fruit are helpful, too. They spread the seeds. Then new fruit trees grow.

In some places, bats visit flowers to feed on nectar, a sweet liquid. As they feed, they scatter pollen from male flower parts to female flower parts. Then the plant can make more fruit and seeds.

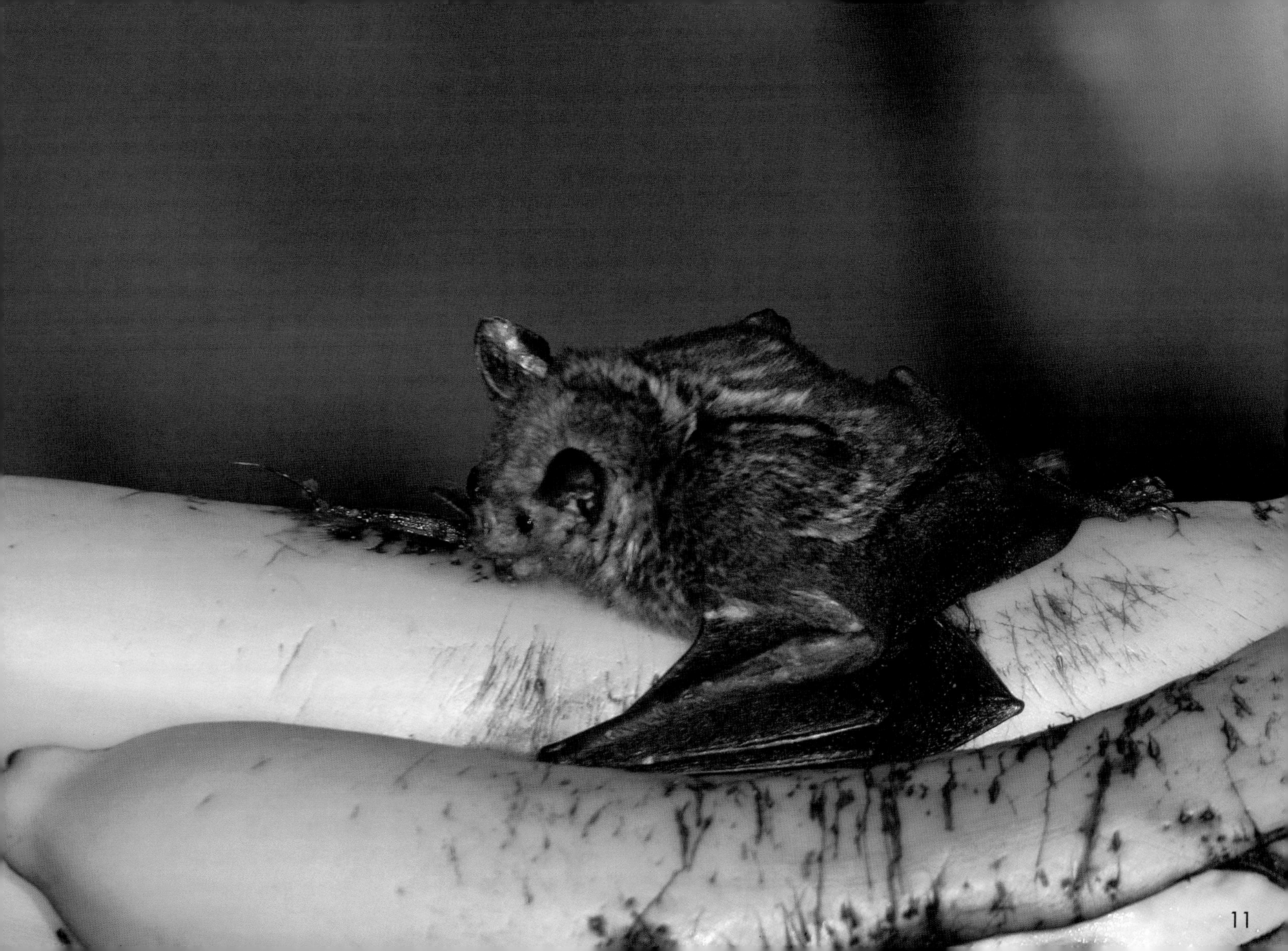

I USED to be AFRAID of BATS,

but now I know . . .

Bats must always watch for danger. Owls and some other animals eat bats. During the day, bats find safe places to rest. Some bats rest in caves. Some rest in trees. Some hide in attics and other places in homes.

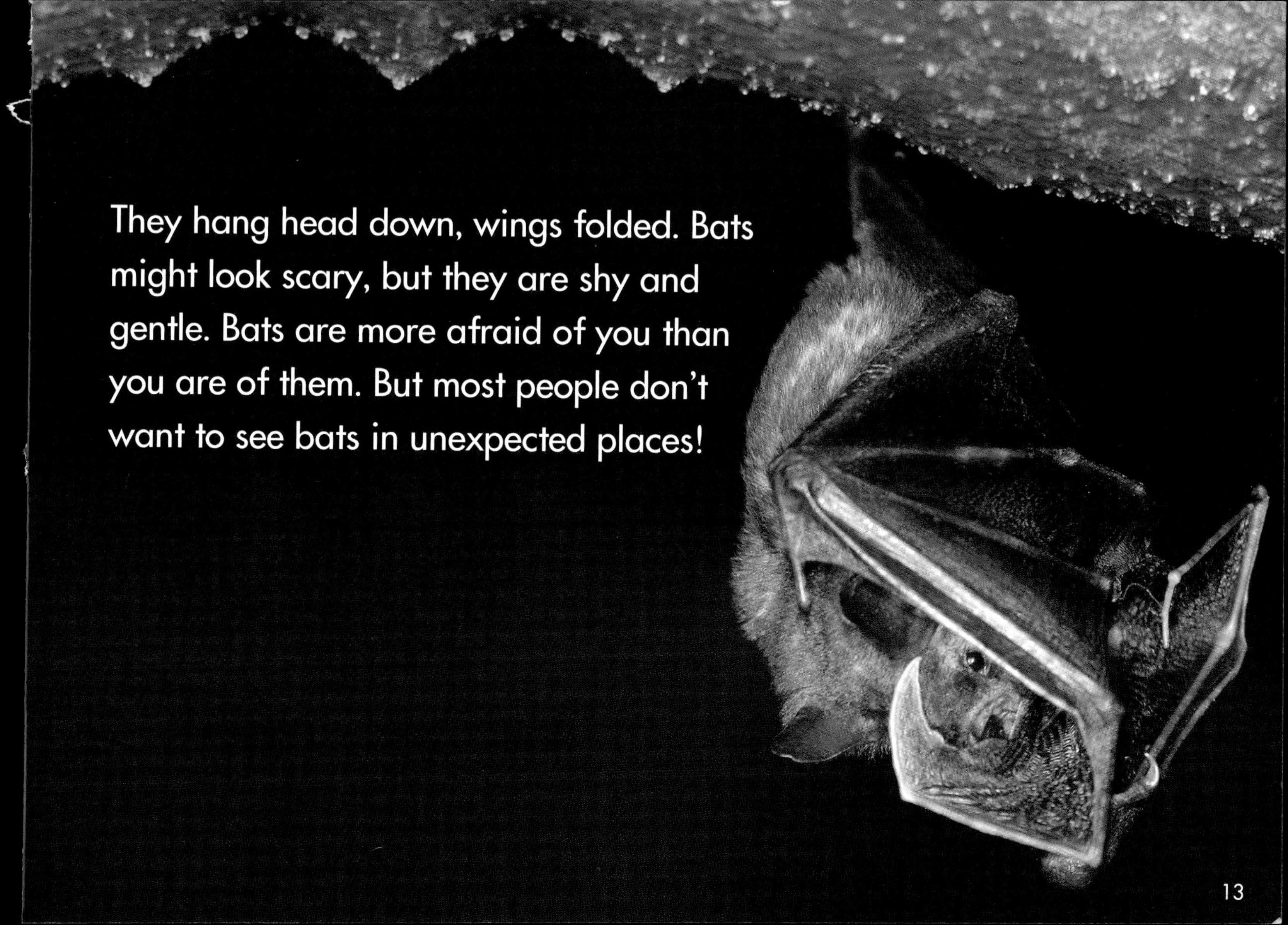

They hang head down, wings folded. Bats might look scary, but they are shy and gentle. Bats are more afraid of you than you are of them. But most people don't want to see bats in unexpected places!

I USED to be AFRAID of BATS, but now I know . . .

Baby bats are born in spring and summer. Most female bats have just one baby a year. This youngster is only four weeks old. It is still with its mother. The baby clings tightly to its mother. She nurses it with her milk. Young bats grow fast.

Animals that have fur and nurse their young are called mammals. Bats are mammals. They are the only mammals that can fly.

I USED to be AFRAID of BATS, but now I know . . .

Bats don't bother people. But if you try to grab a bat, it may bite out of fear. And some bats have rabies, a serious disease. Don't touch a bat, even if it is on the ground. Enjoy these shy and gentle animals by watching them.